Kavin Nataraja

Measuring Alternate Energy Generation via Unity Simulation

AF138610

Kavin Nataraja

Measuring Alternate Energy Generation via Unity Simulation

Presenting a new tool for energy education
M.A.E.G.U.S

LAP LAMBERT Academic Publishing

Impressum / Imprint

Bibliografische Information der Deutschen Nationalbibliothek: Die Deutsche Nationalbibliothek verzeichnet diese Publikation in der Deutschen Nationalbibliografie; detaillierte bibliografische Daten sind im Internet über http://dnb.d-nb.de abrufbar.
Alle in diesem Buch genannten Marken und Produktnamen unterliegen warenzeichen-, marken- oder patentrechtlichem Schutz bzw. sind Warenzeichen oder eingetragene Warenzeichen der jeweiligen Inhaber. Die Wiedergabe von Marken, Produktnamen, Gebrauchsnamen, Handelsnamen, Warenbezeichnungen u.s.w. in diesem Werk berechtigt auch ohne besondere Kennzeichnung nicht zu der Annahme, dass solche Namen im Sinne der Warenzeichen- und Markenschutzgesetzgebung als frei zu betrachten wären und daher von jedermann benutzt werden dürften.

Bibliographic information published by the Deutsche Nationalbibliothek: The Deutsche Nationalbibliothek lists this publication in the Deutsche Nationalbibliografie; detailed bibliographic data are available in the Internet at http://dnb.d-nb.de.
Any brand names and product names mentioned in this book are subject to trademark, brand or patent protection and are trademarks or registered trademarks of their respective holders. The use of brand names, product names, common names, trade names, product descriptions etc. even without a particular marking in this work is in no way to be construed to mean that such names may be regarded as unrestricted in respect of trademark and brand protection legislation and could thus be used by anyone.

Coverbild / Cover image: www.ingimage.com

Verlag / Publisher:
LAP LAMBERT Academic Publishing
ist ein Imprint der / is a trademark of
OmniScriptum GmbH & Co. KG
Heinrich-Böcking-Str. 6-8, 66121 Saarbrücken, Deutschland / Germany
Email: info@lap-publishing.com

Herstellung: siehe letzte Seite /
Printed at: see last page
ISBN: 978-3-659-70678-3

M.A.E.G.U.S

MEASURING ALTERNATE ENERGY GENERATION VIA UNITY SIMULATION

by

Kavin Muhilan Nataraja

ACKNOWLEDGMENTS

I'd like to thank the people who supported this project from its very conception to its completion. Firstly, thanks to Professor David Whittinghill who chose me to work on this brilliant project. Thanks as well to Professor Esteban Garcia and Professor Tim McGraw who served on the committee and provided useful insight and commentary.

Thanks to James He, co-developer of the MAEGUS serious game. We spent many whiteboard sessions arguing, brainstorming and developing the project. I couldn't have asked for a better collegiate partner, roommate and future coworker.

Thanks to William Huynh and Tami Trieu who provided art assets and models for the game for no compensation. Thanks as well to the Colossus Team who worked hard on creating a foundation for the project and providing research value from the MAEGUS pilot study.

Thanks to the Hensersky Family for their prayers and always keeping me in their thoughts, especially during the worst of times. A special thanks to Travis Hensersky for being a great conversationalist and an even better friend.

Finally, thanks to my mother, Malar Nataraja, the best mom in the universe, without whom none of this would have been possible. Thanks for always believing in me and supporting me, no matter the cost.

I put my heart and soul into this, for all of you.

TABLE OF CONTENTS

LIST OF FIGURES

ABSTRACT

Nataraja, Kavin Muhilan. M.S., Purdue University, May 2014. M.A.E.G.U.S: Measuring Alternative Energy Generation via Unity Simulation. Major Professor: David Whittinghill.

This book presents the MAEGUS serious game and a study to determine its efficacy as a pedagogical tool. The MAEGUS serious game teaches sustainable energy concepts through gameplay simulating wind turbines and solar arrays. Players take the role of an energy manager for a city and use realistic data and information visualizations to learn the physical factors of wind and solar energy generation. The MAEGUS serious game study compares game assisted learning to a more traditional teaching method such as reading material in a crossover study, the results of which can inform future serious game development for educational purposes.

CHAPTER 1. INTRODUCTION

Energy awareness is an integral part of today's society. In recent years, the connection between the environment and energy production has become a key factor in energy policy in countries throughout the world. More and more the production of clean green energy has gained importance over continued use of greenhouse gas producing fossil fuels. As a result, curricula at the K-12 and university levels have been reformed to include green energy education. The new curriculum put into place still has gaps for educational tools and this thesis presents a learning module that will fill the gap and provide an enriching educational experience.

1.1 Scope

A serious game called MAEGUS was created in Unity. It has gameplay mechanics similar to SimCity or the Tycoon series of videogames. Players will play through a variety of scenarios where they will be given a budget and various sustainable technologies to meet the energy requirement while operating within the budget. The technologies presented will be limited to wind turbines and solar arrays. Future iterations of the serious game may include non-renewable resources or other sustainable technologies.

The simulation will be given to Purdue University students with interests in sustainable energy related disciplines. The subjects will be administered a post-test pertaining to their knowledge of sustainable energy technologies. The results of thispost-test will be compared with another group of students who did not receive exposure to the MAEGUS serious game but instead received reading material on similar subject material. The short test on sustainable energy concepts will consist of questions such as: "What are the variables that affect wind energy generation?" Retention of concepts after playing the game will be measured by the results of the quiz. If students show equal or higher retention of energy concepts after playing the game as the reading group then the simulation is a successful education tool.

1.2 Significance

Energy literacy is at the forefront of modern education policy. Development of tools to promote energy literacy is a key part of new curriculum design in universities and at the K-12 level. The MAEGUS serious game will optimally become an integral part of energy education curriculum and help teach sustainable energy technology and its benefits to students around the world.

1.3 Statement of Purpose

Energy literacy is an important topic in today's society. The greenhouse gas crisis and the resultant environmental crisis the planet is going through is inextricably related to the use of fossil fuels. Countries worldwide are modifying their energy policies to include more sustainable energy technologies and practices. The MAEGUS serious game will help promote awareness about sustainable energy technologies in high schools and universities by having students solve different energy crisis scenarios with alternative energy choices such as wind turbines and solar arrays. In the process, students will familiarize themselves with the usability and efficiency of these alternative energy systems and become more energy literate as a result.

1.4 Research Question

Can a serious game increase the retention of wind energy and solar energy concepts in college students?

1.5 Assumptions

The assumptions of this study are the following:

- Students will have basic computer skills and will be able to manipulate a Unity game reasonably quickly.
- Students will have access to a quiet and well-lit environment in both versions of the study: the reading portion and the simulation portion.
- Students will have access to computers that can run the Unity simulation and will not crash during simulation.
- Students will be honest in answering the pre-test survey and the post-test survey and will do their best to answer the questions on the concept portion of the test.
- Thirty minutes will be enough time for students to explore the simulation or read the material provided during the testing.

1.6 Limitations

The limitations of this study are the following:

- The study was limited to students classified as college undergraduate and graduate students.
- The study was limited to Purdue University students.
- The researcher cannot control the students possessing prior knowledge of concepts presented in the study and tested in the post-test.
- The sustainable energy concepts presented are limited to wind turbines and solar arrays.

1.7 Delimitations

The delimitations of this study are the following

- The study will not include information on non-renewable energy sources such as coal or oil.

1.8 Definitions of Key Terms

Sustainable Energy - Renewable forms of energy such as wind or solar energy. These forms of energy are sustainable and in most cases considered infinite. "The vast development and deployment of cleaner, more energy efficient technologies will be needed to reduce greenhouse gas emissions" (Kaygusuz 2012, p. 1120).

1.9 Summary

The goal of this thesis is to create an effective supplement to energy education curricula. It will be created in Unity and will be tested with college students to determine its efficacy as a pedagogical application. In order to create a valid simulation as well as a survey to test it a literature review was conducted which is detailed in the next section.

CHAPTER 2. LITERATURE REVIEW

Green energy is a large constantly evolving topic covering production methods that are being refined and revised for optimization. This literature review presents the need for the MAEGUS serious game by identifying the gap in energy education curriculum. The first section of this paper will provide insight into green energy and the methods of alternate energy production simulated for testing: wind turbines and photovoltaic arrays. Existing alternate energy education policies and methods will also be examined to show the need for an educational tool. Additionally, previous serious games research will be discussed. Finally, algorithms and formulae for the physical energy simulation will be provided from the source material.

2.1 Green Energy

Modern countries are defined as modern by their nationwide access to energy and poverty of developing countries is calculated by the same measure (Kaygusuz, 2012). Energy poverty is a lack of access to electricity or modern sources of energy or the reliance on inefficient and finite energy sources such as biomass. Biomass energy is derived from the burning of materials such as firewood, charcoal and dung (Kaygusuz, 2012). While biomass is suitable for activities such as cooking, modern services require more stable and sustainable sources of electricity. In the domestic environment, refrigeration and electrical lights provide convenience while modern electric-stoves that cut down on fumes from burning biomass can prevent health issues. On a larger scale, hospitals use refrigeration for sterilization and storage and manufacturing industries require electricity to power plants (Kaygusuz, 2012).

In 2012, an article in the *Renewable and Sustainable Energy Reviews* made a case for reforming energy policy to help developing countries of the world. The article shows that close to 1.4 billion people live without electricity in the modern world and this number will only increase in the next 20 years if current energy policies are left unchanged (Kaygusuz, 2012).

What remains constant throughout the projected years is increasing consumption of fossil fuels. Currently 21% of energy production comes from gas and coal consumption and this number is projected to rise to 23% by 2030 (Kaygusuz, 2012). The oil trade fuels the world economy and as a result many developing countries with naturally occurring fossil fuel resources choose to export them while relying on biomass fuel resources in their own homes, which propagates the lack of electricity in developing nations (Kaygusuz, 2012).

An assumption drawn from the continued growing use of coal and oil is that CO_2 emissions will also increase in the future. Kaygusuz (2012) postulates that there will be up to a 55% increase in CO_2 emissions in the next two decades. The burning of fossil fuels is the largest contributor of greenhouse gases and is a primary cause of the deterioration of the environment (Chen, Huang & Liu, 2013). The burning of fossil fuels and production of CO_2 isn't just restricted to developing countries. Modern nations consume even more energy using fossil fuels due to new technological developments (Kaygusuz, 2010; Kaygusuz, 2012). As India and China are expanding economically and other countries increase in population so does energy consumption and as a result the demand and subsequent consumption of fossil fuels increases (Kaygusuz, 2010).

While a variety of policies have been passed in response to the global warming crisis, the universal energy industry is broad and no one policy can effectively regulate the entire system (Kaygusuz, 2012). Even so there is a large amount of research being conducted into cleaner, greener energy production.

In the area of coal and oil consumption, methods are being researched to regulate carbon emission while retaining efficient energy generation. Carbon Capture System (CCS) is a method of capturing carbon while coal and natural gas is being burned. This process optimally can completely eliminate carbon emission from these processes (Kaygusuz, 2012). Research though still needs to be expanded in this and other processes. Until processes like CCS become reliable and widespread, alternative methods of energy production must be adopted.

The alternative to fossil fuels is the harnessing of renewable or sustainable energy sources for energy generation. Kaygusuz (2012) wrote that "access to sustainable energy sources should form a central component of broader development strategies" (p.1119). Sustainable energy sources that provide energy exempt of CO_2 emissions and are reliable will alleviate the problem of energy poverty while reducing the impact of the greenhouse gas crisis.

Governments around the world have begun emphasizing research into these alternative energy sources to provide relief for the environment as well as to avoid crisis as a result of the inevitable fossil fuel shortage (Chen et al., 2013; Garg & Kandpal, 1996; Kaygusuz, 2010). These alternate energy sources include wind turbines, hydroelectric generators, solar panels, biomass generators and geothermal plants (Kaygusuz, 2010).

In 2007, 3% of all electrical energy worldwide was created by sustainable energy resources, with the exception of hydroelectricity. By 2030 this percentage should double to 6% with the largest portion of the increase attributed to wind turbines (Kaygusuz, 2010). Wind energy is the fastest growing sustainable energy thanks to government incentives that fund wind farms. The United States, Denmark, Turkey, China, Egypt and other nations all have policies that promote wind technology (Saidur, 2010). Solar energy is another major sustainable resource and a case study in 2011 calculated that solar panels could provide 16% of the total annual electricity consumption of Taiwan (Yue & Huang, 2011).

Yet despite the efforts put forth by governments worldwide, there has been a widespread lack of adoption of alternate energy technologies for a number of reasons. First, without a central infrastructure to guide sustainable energy education, developing countries have opted to continue using fossil fuels or have adopted inefficient or expensive technologies (Acikgoz, 2011). Second, without proper education in maintenance or repair of sustainable energy technologies, some countries created a combination of technologies, that were ultimately abandoned by end-users who didn't know how to properly utilize the benefits of the technology (Acikgoz, 2011).

Both problems stem from a lack of energy education. Clearly there is a need for alternate energy adoption. In order to increase awareness of alternate energy technologies education policies have been revised to include sustainable energy education (Acikgoz, 2011; Garg & Kandpal, 1996). Following is a brief look at existing energy education policies and what researchers have deemed important in promoting alternate energy technologies.

2.2 Energy Education

Sustainable energy education seeks to promote the adoption of sustainable energy technologies by increasing knowledge of alternate energy solutions at the K-12 and university levels (Garg & Kandpal, 1996). The oil crisis of 1973 led to a boom in awareness of the finite quality of fossil fuels. Fossil fuels changed from a commodity to a scarce dwindling resource leading to an increase in general knowledge of oil industry and nonrenewable resources (Kandpal & Garg, 1999).

A person at the end of an energy education program is optimally "energy-literate". Dewaters and Powers (2011) defined energy literacy as more than just content knowledge. Energy literacy includes "a 'citizenship understanding' of energy that encompasses affective and behavioral aspects" (Dewaters & Powers, 2011, p.1700). An energy-literate person is not just aware of the environmental impact of energy, but is also inclined towards energy conversation. This means that energy education ultimately produces an individual who has an active comprehension of energy and not a passive understanding that most curricula impart (Acikgoz, 2011; Dewaters & Powers, 2011).

The lack of enthusiasm to adopt sustainable energy technologies is directly correlated to a lack of energy literacy (Dewaters, 2011). By promoting strategies of energy saving, governments can focus on developing new technologies and methods of energy conservation instead of relying on acquiring an abundance of fossil fuels or other energy sources. This will lead to a more active stance on adopting sustainable energies and help developing countries and first world countries deal with energy poverty and fossil fuel depletion (Dewaters, 2011).

Studies of energy education still remain scarce. Since energy is studied in other curricula such as engineering and physics it has been slow in gaining status as its own separate discipline. Where energy is treated as its own separate discipline mainly is in independent courses or minors, created with funding from research grants or government funding (Stone, 2011). Energy is usually studied as an additional discipline in mechanical engineering or civil engineering (Acikgoz, 2011; Stone, 2011).

The need for energy education to be recognized as an independent discipline is important as it will lead to an increase in energy literacy (Acikgoz, 2011). In order for energy education to become its own discipline, a framework and goal must be developed. Jennings (2001) wrote that teachers and education would be key in promoting Ecologically Sustainable Development (ESD) which is defined as:

9

Development which meets the needs of the present without compromising the ability of future generations to meet their own needs and, improving the quality of life while living within the carrying capacity of supporting ecosystems. (Jennings, 2001, p.114)

The acceptance of the concept of Ecologically Sustainable Development is the main goal of energy education. In 1992, the United Nations Commission on Environment Development passed Agenda 21. Agenda 21 defined precisely the role of energy education in a society striving to adopt sustainable energy technologies:

- To promote environmental and ethical awareness, values, attitudes, skills and behaviour needed for sustainable development;
- To build capacity for nations to develop Agenda 21 action programs;
- To train more scientists and engineers with an understanding of ESD and the technology for sustainable development;
- To re-orient all levels of education towards ESD (Jennings, 2001, p.115).

Since Agenda 21's passing, developing countries have formed their own agendas and policies on energy education and ESD. While their names for their frameworks are different, they share many similarities and the different programs have overlapping definitions of what makes an optimal energy education program. Kandpal and Garg (1999) created a list of the desirable features of an energy-education programs that overlapped with other researchers' optimal programs including that they should:

1) include all energy resources (renewable and non-renewable) (Acikgoz, 2011; Kandpal & Garg, 1999)
2) cover all aspects of energy technologies such as:
    a)   resource assessment
    b)   technology
    c)   economics and energetics

10

d)   sociocultural issues, and

e)   ecological and environmental impacts

3) be developed for different educational levels and for different audiences (Acikgoz, 2011; Kandpal & Garg, 1999)

4) be consistent at the local, national, regional and international levels.

5) be flexible and dynamic.

6) ensure employment/self-employment for students and have a direct link with job requirements and responsibilities required in the area of energy.

7) be compatible with global efforts (Kandpal & Garg, 1999).

The requirements requested above will be integrated into the final simulation's design. The final simulation created will be administered to university level subjects and will have to balance the benefits and consequences of different renewable resources.

Following are some examples of education policy and curriculum design, categorized by country, that provided insight into designing the final simulation.

2.2.1 Taiwan

In 2009, Taiwan created "Energy Saving and Carbon Reduction", an action plan to promote a sustainable energy policy. The goal of the plan was not only to increase awareness and knowledge of low carbon energy but also to embed the "energy-related comprehension into low-carbon and energy saving attitudes for citizens" (Chen et al., 2013, p.397). It is important that energy education doesn't just provide comprehensive knowledge, but that it also makes individuals more adaptive to new sustainable energy technologies. This goal runs parallel to Agenda 21's first goal: "Promote environmental and ethical awareness, values, attitudes, skills and behavior needed for sustainable development" (Jennings, 2001, p.115). Consistent with goals expressed in "Energy Saving and Carbon Reduction", Taiwan's Ministry of Education and the National Science Council created the "National Science and Technology Program" and started research on curriculum redesign as well as assessments to test energy literacy (Chen et al., 2013).

Kuan Li-Chen, Su-Hun Huang and Shiang-Yao Liu (2013) developed a survey questionnaire during the "National Science and Technology Program". The purpose of the survey was to determine the dimensions experts found most important when trying to improve energy literacy. The survey was given to experts defined by the authors as fellow researchers in the "National Science and Technology Program" or involved in the "Energy Technology Education Center" (Chen et al., 2013). Thirty four percent of the experts were professors in environmental related disciplines and 66% of the survey subjects were teachers in schools partnered with the "Energy Technology Education Center" (Chen et al., 2013). The authors found that "civic responsibility for a sustainable society" was the highest priority among test results. Other dimensions included "low-carbon lifestyle", "energy concepts" and "reasoning of energy issues" (Chen et al., 2013, p.398). Of note was that "possessing a systematic understanding about energy" was the lowest priority indicator (Chen et al., 2013, p.401). A conclusion can be drawn from this is that experts didn't believe that citizens needed a thorough understanding of energy to be energy literate, but did need to be aware of sustainable energies and their overall benefit to a community. The authors published their research as an energy education framework to be implemented in school curricula in Taiwan (Chen et al., 2013).

Taking into account the dimensions discussed above, the simulation created will be as realistic as possible but the game mechanics will be intuitive so users won't have to learn the technical details of energy production and management but will be engaged in crucial decision-making that will impact their community and the budget given to them.

2.2.2 Turkey

Acikgoz (2011) writes that energy education should be introduced at the grade school level and universities should also offer full degrees that end with a diploma in energy related fields. A training course for employed engineers or engineers looking for employment could also help bolster careers with energy literacy. Turkey has put into place many incentives for sustainable technology research and development. The adoption of numerous Five-Year Development Plans by Turkey have led to large scale government policy changes focusing on different parts of Turkey's infrastructure (Acikgoz, 2011). The eighth Five-Year Development Plan focused on education development and increase communication technology and curriculum in schools. The latest Five-Year Development Plan, the ninth one lasting from 2007-2013, focuses on meeting the increasing demand for energy. As laid down by the preceding Five-Year Development Plan, new curriculum programs are being developed to increase the infrastructure of Turkey's educational system. The new curriculum introduced in the ninth development plan focus on sustainable technologies and energy education at various levels (Acikgoz, 2011). Turkey's goal in its latest iteration of the Five-Year Development Plan is to help ease the demand for sustainable technology and Acikgoz (2011) writes that the renewable energy industry and Turkey's interaction with it will be greatly influenced by education.

The focus of the energy education curriculum is on "the most important renewable sources" listed as hydropower, wind, solar, geothermal and biomass (Acikgoz, 2011, p. 608). Turkey's emphasis on these sources comes from its rank as seventh in the world in terms of geothermal production potential, and Turkey has 1% of the total world hydroelectric energy production (Acikgoz, 2011). Traditional university courses will not be able to keep up with rapidly developing sustainable energy technologies. Instead short training courses for college students and graduates will be crucial in promoting awareness of Turkey's sustainable technology needs and what the county has already accomplished (Acikgoz, 2011). An important part of the education process will also be modern educational technology such as e-Learning courses. Modern instruments that utilize software or the internet are appealing to teachers and students (Acikgoz, 2011). The inclusion of digital learning materials such as PDFs and image files or videos accessed through a PC can make web-based training units possible. Even flash files are being implemented in new courses at universities and at the school level (Acikgoz, 2011).

This shows that there is a place for the final simulation. A Unity-based simulation could be used in web-based training units and is also an attractive alternative for educators and students. The final simulation could be an integral part of a new e-learning approach to energy education.

2.2.3 United States

In the United States, energy education is still a sub-discipline and relegated to a minor or addition to normal course workloads. Universities in the United States are focusing on adding energy education to their curriculum with the incentive of government funded grants (Stone, 2011). At the Colorado School of Mines, a new energy minor curriculum is in development with funding from a NSF grant. Alongside the new curriculum a student energy club and summer research program are being developed (Stone, 2011). The minor presents courses for students who have technical backgrounds in mathematics and physics and provide hands-on in class exercises as opposed to survey style lectures (Stone, 2011). Students responded favorably to course work that had them working on hands-on technical problems. Community service projects and field trips were other activities that students were attracted to (Stone, 2011).

Stone (2011) provides examples of other United States schools that focus on lecture based energy education courses. The uniqueness of the Colorado School of Mines energy program shows that hands-on activities and exercises can be as educational as lectures while more favorably viewed by potential students and teachers (Stone, 2011). The final simulation will be a hands-on part of educational programs and will allow students to participate actively in a simulated environment.

2.2.4 Summary

The final Unity simulation reflects the developing energy education curriculum design in many countries: a hands on all-inclusive instructional program that is relevant at multiple educational levels and is easily integrated into existing curriculum. The final product of this paper will be the kind of e-learning experience students look forward to and benefit from as much as a lecture based session. Researchers and governments have already established the need for more sustainable energy and energy literacy. The solution to the need must be innovative and engaging to be adopted by schools and universities.

2.3 Serious Games

Serious games is a genre of games that goes beyond just pure entertainment. As de Fritas wrote in 2012, serious games serve to improve important skills such as "problem-solving, decision-making, inquiry, multitasking, collaboration and creativity" (de Fritas et al., 2012, p.289). Serious games can immerse students in scenarios and environments they normally couldn't experience or interact with. These serious game experiences have been reviewed favorably by students and faculty when they are included in educational curricula (Stone, 2011; Darling et al., 2008).

2.3.1 Game Assisted Learning

Game assisted learning is defined as "the outcome of integrating effective learning principles into game environments for the purpose of utilizing engaging elements of games as a means for improving the quality of education" (Wu, Chiou, Kao, Alex Hu & Huang, 2012, p.1156).The relative affordability of digital game devices and the favorable reaction from educators and students has led to the development of pedagogical game-based applications (Wu et al., 2012). Wu and his fellow researchers (2012) conducted a meta-review of a number of game assisted learning papers that reviewed periods of game assisted learning research. The papers covered game studies conducted throughout 1963 to 2007 and investigated the application and results of instructional game design research. The concepts that were gamified in the studies included math, physics, business, social sciences and engineering. Wu found that while serious games weren't applicable in all instructional environments they had a significantly positive effect on student's learning outcomes, especially in the subject area of math (Wu et al., 2012).

Game assisted learning is still nascent and suffers from a lack of distinct methodology (Mayer 2012). However two important concepts of game assisted learning are accountability and responsibility. Accountability ensures users have their expectations met by serious games in terms of content and experience. Developers have the responsibility of actively and effectively reaching their goals of education or training using their software. MAEGUS was developed with these key concepts in mind (Mayer 2012).

While there have been numerous instances of gamification in course curriculum, the major inspiration for MAEGUS and its testing methodology comes from the Racing Academy project (Darling et al., 2008). Racing Academy was developed to help UK higher education students develop an understanding of the engineering concepts in vehicle dynamics using computer gaming (Darling et al., 2008). A pre-test was administered to a class of approximately 160 students a week before exposure to the game began. The pre-test tested engineering concepts that would be presented in the game. After playing the game students were given an identical post-test which also tested engineering concepts and significant improvement was found in the second round of testing (Darling et al., 2008). MAEGUS will follow a similar design by giving students a pre-test on sustainable energy concepts and then compare these results with results on a post-test after the MAEGUS game has been played to determine how it affects learning outcomes.

2.3.2 Flow Theory

A major factor of game enhanced learning is related to the theory of game flow. Flow theory, introduced by Csikszentmihalyi in 1975, deals with player engagement and the balance between boredom and anxiety in games (Csikszentmihalyi, 1975).

Flow is "a state of complete absorption or engagement in an activity and refers to the optimal experience." (Kiili, 2012). When a person enters this state of optimal experience, they focus singularly on the task at hand and ignore outside factors. The flow theory has been applied to multiple fields of research but its acceptance in game research is important for the state of game assisted learning. By utilizing the theory of game flow, researchers can develop pedagogical applications that engage a user to a point that they do not realize they are learning as they are playing a game (Kiili, 2012). The goal of MAEGUS is to provide an engaging experience that will convince students to play the serious game for entertainment purposes while they subconsciously achieve learning outcomes.

2.4 Simulation

In order to gamify sustainable energy technologies, MAEGUS realistically simulates energy using existing formulae and scientific factors discussed in the next section. The product of this thesis will focus on wind turbine simulation and solar array simulation.

2.4.1 Wind Turbines

A number of environmental factors as well as technical specifications affect a wind turbines' energy output. In the final simulation, the factors of energy generation that go into energy generation and energy management are the important concepts that serve as learning outcomes (Kandpal & Garg, 1999).

The independent variables in the wind turbine formula are the swept area of the rotor blades (A), air density (p), the wind speed (v) and the power coefficient of the generator (Cp) with the dependent variable being power generated (P) (Belu & Koracin, 2012, Foley & Gutowski 2008). The wind turbine formula used in MAEGUS is shown in Figure 2.1 and is the summation of all wind turbines in the game:

$$\sum \frac{1}{2} pAv^3 C$$

Figure 2.1. Wind Energy Formula

The length of the wind turbine blades are crucial to determining the swept area. In order to find the swept area, the circle area formula is used: πr^2 (Belu & Koracin, 2012). Air density is determined by the wind turbine elevation and the power coefficient is the generator's mechanical efficiency which is governed by the Betz limit. The average is around 40% (Belu & Koracin, 2012). Wind speed will be simulated by weather patterns aggregated from an online database.

Other factors to consider are wind direction and distance between turbines. A study was conducted in 2012 that showed that wind turbines suffer minimal loss of energy when wind direction deviates up to 90 degrees from the horizontal axis. If the wind's angle of incidence is above 90 degrees to either side of the wind turbine, there will be a loss of energy generation (Chen, Wang, Liu, Chen, Li & Guan, 2012). A wake effect exists if wind turbines are placed too close together. However, because most modern wind turbines automatically rotate their rotor to face the optimal wind direction, wind direction wasn't simulated in the build of MAEGUS presented in this paper. Studies show that the wake effect can be removed if the turbines are placed at least 500 meters away from each other, and each unit is subject to the same average wind speed (Chen et al., 2012). In game, this is represented by the wind turbines taking up a large unit of space where nothing else can be built adjacent to them. This spacing simulates the need for consideration of wake effect space.

2.4.2 Solar Arrays

Also called photovoltaic trackers, solar arrays are scalable in size by the number of panels in an array (Koussa, Cheknane, Hadji, Haddadi & Noureddine, 2011). Solar arrays are measured by the estimated wattage of the panels and the number of panels in an array (Koussa et al., 2011). Another factor of solar energy generation is the material of the panels. There is a wide variety of materials with varying efficiencies of photovoltaic absorption (Green, 2013). The angle of incidence also plays a key role in how much sunlight a solar array gathers. Throughout the day solar arrays tilt to optimally gather as much sunlight as they possibly can. Research has been conducted on formulas to measure the optimal tilt of solar arrays based on the sun's location as well as weather patterns (Koussa et al., 2011; Mehleri, Zervas, Sarimveis, Palyvos & Markatos, 2010).

2.4.3 Conclusion

There are a number of factors to take into account when programming the physics of both wind turbines and solar panels. This simulation only focuses on these two sustainable resources. Plans for other renewable and nonrenewable resources can be considered for future studies. How the simulation will accurately implement the physics discussed above and tie them together with game play mechanics is discussed in the Methodology section.

CHAPTER 3. METHODOLOGY

In order to measure retention of energy concepts, test scores on the MAEGUS study survey were evaluated. Participants were split into two groups: group 1 played the MAEGUS serious game while group 2 reviewed written material covering sustainable energy concepts. At the end of the review session an identical post-test survey on sustainable energy concepts was given to both groups. The independent variable of the study was the method participants used to learn sustainable energy concepts. The post-test was administered to determine retention of sustainable energy concepts after exposure to the two different educational methods. The dependent variable of this study was the results of the post-test of both groups.

The hypotheses for the study follow:

H_0: There is no difference between test results of a group given a serious game versus a group given reading material.

H_a: There is a difference between test results of a group given a serious game versus a group given reading material.

Following is more information about the MAEGUS serious game, the reading material, the assessment created for the study and the testing procedure itself.

3.1 <u>MAEGUS Gameplay</u>

3.1.1 Narrative

MAEGUS is a turn-based simulation game that takes place in a far future setting where energy resources have run out. Civilization has collapsed and the player takes the role of the MAEGUS a wandering nomad who discovers the last city in existence. The city has a bare supply of energy and its elder, the Sage, asks the player to help him build new energy generators using plans for wind turbines and solar arrays he has salvaged. The MAEGUS has the ability to visualize the world as hexes and can also see the energy factors of each hex in the world. As the MAEGUS, the player builds wind turbines and solar arrays in these hexes and must interpret the energy factors in each location to optimize their energy output. Narrative, which has been shown to increase engagement in roleplaying oriented games, was added to promote flow and interest in MAEGUS (Dickey, 2006).

Figure 3.1. The MAEGUS interface

3.1.2 Game World

The MAEGUS game world is split into hexes, which are hexagonal space identifiers. Each hex represents 500 square meters in game and has energy factors such as wind velocity and solar concentration tied to them. These hexes are further organized into sixteen regions and a stochastic weather model was developed to randomly seed each region based on a weather dataset (Wilks & Wilby, 1999). The weather dataset used is realistic weather data taken from the National Renewable Energy website. The game simulates over 20,000 possible data points of wind data and solar energy from the state of Indiana, the chosen region that the prototype level simulates (NREL). When a player hovers over or clicks on a hex they are presented with the energy generation formulas for wind energy and solar energy as well as a comparison of the two. Data visualizations were created to help the players understand the energy factors in each region and how to best adapt their build strategy accordingly.

Figure 3.2. MAEGUS with Data Vis Windows

The random seed stores within each region an overall weather model and each turn the weather factors in the hexes within that region vary depending on the seed. As such, MAEGUS game sessions vary greatly in terms of how the player can generate the most energy. Some sessions solar energy is optimal because of more stable solar concentration while other game sessions wind energy is more effective because of higher wind speeds. This emphasis on the realistic external weather factors of energy generation were designed to be engaging as a gameplay mechanic and to function as an important learning outcome.

3.1.3 Challenges

Players must be careful about how they build not only because of the factors of energy generation in hexes but because they are also restricted by the amount of space they can build in and by their funds. At the beginning of the game, the player's city has a limited build radius around it. Players can only place their energy generators in the radius around their city. They are also given limited funds at the beginning of the game that restricts how many energy generators they can build. Additionally, wind turbines and solar arrays differ in how much they cost and wind turbines cannot have anything in adjacent hexes to them. After building an energy generator, players cannot move or destroy their generators. The limitations of space and funds provide challenges to the player to carefully analyze their choices before making them to generate as much energy as possible.

Figure 3.3. MAEGUS City Level 1

The city hex is the player's lifeline. The city provides the players with energy goals that they must reach using their energy generators. By generating energy and reaching an energy goal, the player's city levels up, visually growing bigger, providing more funds and creating a larger build radius. Every time a player reaches an energy goal the limitations on their gameplay is decreased and a new energy goal is provided which pushes players to build bigger and better energy generators.

MAEGUS in the test build consisted of 20 turns. Turns represent a full day of energy generation for wind turbines and solar arrays. At the beginning of every turn, players are given a base amount of funds from their city and a bonus amount based on how close they are to their next energy goal, as a reward for progressing. At the end of every turn, MAEGUS calculates the amount of energy each wind turbine and solar array produces in kilowatt hours (kWh) and adds it to the player's total energy generated. The turns were also a time limit that forced players to strategize and use their funds to their maximum potential.

Funds could also be used on upgrades for energy generators. The two internal factors of wind energy, blade length and generator efficiency, could be upgraded for wind turbines while solar arrays could have their estimated wattage and material type upgraded. Upgrades increased in price exponentially and each of these internal factors were also represented in the energy formulas. Players in the end game could decide to prioritize upgrading their energy generators instead of building more as their finite build space ran out. Upgrades were also represented realistically in the energy generation formula and had long term impacts on an energy generator's efficiency.

A final incentive for players is that at the end of the game, their score is judged and added to a high score list if high enough. The high score system is an extra incentive to motivate players to generate as much energy as possible to top their peer's scores.

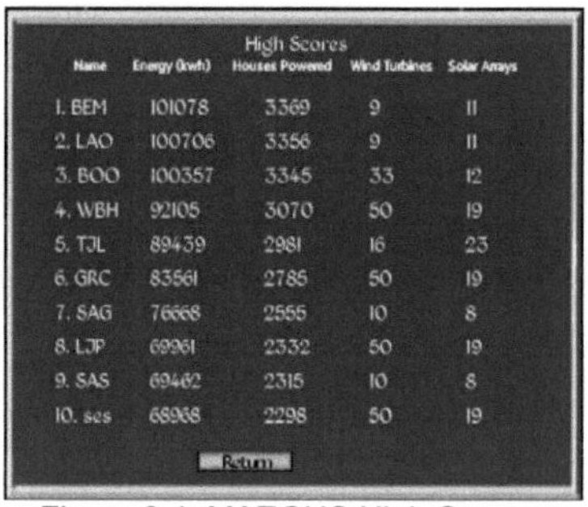

High Scores

Name	Energy (kwh)	Houses Powered	Wind Turbines	Solar Arrays
1. BEM	101078	3369	9	11
2. LAO	100706	3356	9	11
3. BOO	100357	3345	33	12
4. WBH	92105	3070	50	19
5. TJL	89439	2981	16	23
6. GRC	83561	2785	50	19
7. SAG	76668	2555	10	8
8. LJP	69961	2332	50	19
9. SAS	69462	2315	10	8
10. ses	68968	2298	50	19

Return

Figure 3.4. MAEGUS High Scores

3.2 Reading Material

The reading material used in the MAEGUS study was composited from three information pamphlets from the NREL website (NREL). The three pamphlets were dissected for the most relevant material to the study. The first pamphlet from 2001 was titled *Nonrenewable Energy: An Overview* and provides an overview of nonrenewable energy and sustainable energy sources. The overview section and section on wind energy and solar energy were taken from this paper. The second pamphlet from 2008, *30% Wind Energy by 2030*, promoted having wind energy provide 30% of the total energy produced in the United States by 2030. The portion on the energy generation factors was used as reading material. The final pamphlet, *Get Your Power From the Sun*, had information on solar arrays and their factors of energy generation and efficiency.

These pamphlets were composited into a comprehensive ten page packet of information on sustainable energy with a focus on wind power and solar power. This reading packet was used in the reading group to compare MAEGUS to traditional material used in classrooms or as study material.

3.3 Assessment

Development of MAEGUS took precedence during this thesis. In order to test its efficacy as an educational tool, a comparison of test scores with reading material was decided on. Since development on MAEGUS took priority, the researcher did not have the time or experience to develop a scale specifically for the study. Instead, the researcher used affective five point Likert scale questions from the MAEGUS pilot study conducted in 2013 (Nataraja, Whittinghill & Head, 2013) and combined them with cognitive multiple choice questions from DeWaters' Energy Literacy Survey (Dewaters & Powers, 2008) as well as created custom questions along the lines of those asked in the Energy Literacy Survey.

In the MAEGUS pilot study conducted in 2013, a prototype build of MAEGUS was tested with undergraduate students in Purdue's College of Technology to determine if MAEGUS had any significant impact on student's attitudes towards sustainable energy concepts (Nataraja, Head & Whittinghill, 2014). In the study, players were given a pre-test with five likert scale questions that tested their attitude and confidence towards sustainable energy, asking them to rate how much they agreed with statements such as: "I am interested in sustainable energy." (Nataraja, Head & Whittinghill, 2014). After a gameplay session with MAEGUS an identical post-test was

27

administered and the results of the pre-test and post-test were compared. The researchers found a significant positive increase in participant's attitudes after playing MAEGUS among other results. The MAEGUS study uses these affective likert scale questions to measure confidence and interest in participants.

The scale to measure energy literacy was adapted from Dewaters' Energy Literacy Survey (ELS). Originally tested with students at the high school level, Dewaters developed the ELS to measure energy literacy (DeWaters & Powers, 2008). The Energy Literacy Survey consists of 69 questions split into the three categories of behavioral, affective and cognitive questions. The MAEGUS study used four multiple choice questions from the Energy Literacy Survey's cognitive question section: questions 37, 41, 43 and 44 (DeWaters & Powers, 2008). Appropriate for general knowledge about energy, these questions were used as the first four questions in the cognitive section of the MAEGUS study. The remaining six cognitive questions were created by the researcher specifically to test sustainable energy concepts presented in the MAEGUS game and the reading material such as: "Wind Turbines derive energy by...".

The final resulting fifteen question MAEGUS Study survey can be found in the Appendix. The first five likert scale questions are affective questions while the final ten are multiple choice cognitive questions. This study was administered to all participants in the MAEGUS study, both in the reading group and the game group, and the test scores of the two groups were processed with a comparative analysis to test the hypotheses of the thesis.

3.4 Procedure

3.4.1 Sample Population

After receiving IRB approval for the human testing portion of the MAEGUS study, testing began. The maximum sampling size for the MAEGUS study was 50 participants who were split into two equally sized groups of 25. The sample population was a convenience sample from Purdue University students of any major or classification. Flyers, departmental emails and word of mouth was used to recruit participants for the study. A majority of test participants were College of Technology students who were aware of the testing because of the Games Research Lab's work.

3.4.2 Testing Environment

The MAEGUS study was conducted in Purdue University's Games Research Lab in the Computer Graphics Technology department. The Games Research Lab is an enclosed space with a Windows computer and speakers. Participants were allowed to walk-in to the Games Research Lab in between testing sessions to either participate or set up an appointment to test. The room was closed during testing and locked from the inside so passersby couldn't see or enter the room while testing was in progress. The MAEGUS study was conducted in this quiet environment where participants could concentrate on the study.

3.4.3 Testing Process

Before testing began, the research team randomized the order that participants would be placed into the two groups. Group 1 was designated as the game group and Group 2 was designated as the reading group. Both groups had an equal number of participants, 25 in each. When participants arrived for the testing session, the researcher gave them the IRB approved consent form and obtained their signature as well as provided a copy of the form for the participants' records. Participants were allowed to leave at any time and were assured the study was completely voluntary. Participants were informed of the existence of two different groups for the study, the game group and the reading group, before their consent was attained. Participants were not told what group they were in until after the consent form was signed and testing began. At that time, the researcher consulted the randomized order to determine which group the participant was queued for.

After assigning the participant to a group the testing began. Participants in group 1 were shown a short tutorial video of the controls of the MAEGUS serious game. After watching the video, the researcher started the MAEGUS gameplay session. Participants played the MAEGUS serious game and were timed during gameplay. The MAEGUS test build has 20 turns and if the participants finished their turns before 20 minutes the timer was paused and the post-test survey was administered. If participants reached 20 minutes and hadn't finished the game the researcher asked the participants to stop building energy generators and to just cycle through their remaining turns to see how much energy they produced with the buildings they created. The post-test survey was then administered.

Participants in Group 2 were asked to read the information packet described in Section 3.2. They were given 20 minutes to read the material and were timed. If participants didn't finish within 20 minutes they were asked to stop reading, the information packet was returned to the researcher and the post-test survey was administered. If participants finished early, their time was noted and the information packet was retrieved by the researcher before the post-test survey was administered.

Participants were given as much time as needed to complete the post-test survey, typically a 5 minute task. Afterwards they were debriefed about the study. The post-test survey and consent forms were stored in secure file lockers and later the surveys were graded and the results were recorded for data analysis.

3.5 <u>Summary</u>

The MAEGUS study tests whether or not a goal-based serious game can motivate students to learn sustainable energy concepts as effectively as reading material can. The MAEGUS study was conducted at Purdue University with a convenience sample of 50 college students, split into two equal groups of 25. These groups were exposed to different educational methods and the results of identical post-tests were compared to determine if there is a significant difference between serious games and reading material in regards to teaching sustainable energy concepts. Following is the Results section where the data from the MAEGUS study is statistically analyzed.

CHAPTER 4. RESULTS

The MAEGUS study survey consists of 15 questions: 5 likert scale attitudinal questions and 10 multiple choice cognitive questions. The likert scale questions used a 5-point likert scale and were scored on a 1-5 scale with 1 being highly negative attitude and 5 being highly positive attitude. The cognitive questions were scored out of 10. The scores on each attitudinal question and the cognitive scores were grouped into their respective group for group 1 and group 2. The scores for each group were then analyzed. For the likert scale questions the responses are discrete and so a Mann-Whitney U test is used to compare if the two groups have the same mean alongside the independent two-sample t-test to see if there is significance. Independent two-sample t-test was used on the cognitive scores. An alpha of .05 and a confidence level of 95% was used in the t-test analyses. Following are the results of the analyses. In the analysis, group 1 represents the serious game group while group 2 represents the reading group.

4.1 <u>Attitudinal Likert Scale Analysis</u>

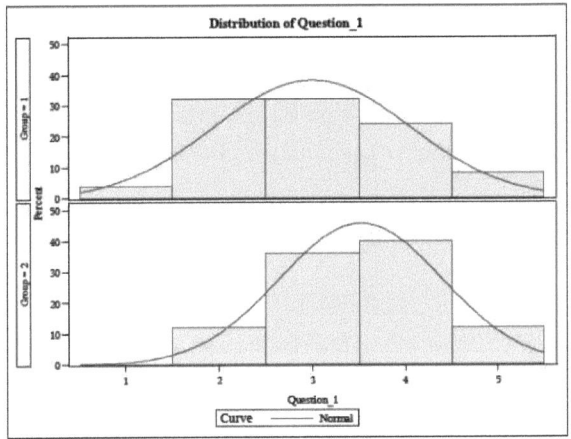

Figure 4.1 Distribution of Likert Scale Question 1

Question 1 of the MAEGUS study survey is "I have an in-depth knowledge of sustainable energy" with 1 meaning the participant strongly disagreed with the statement and 5 being the participant strongly agreed with the statement. The Mann Whitney U test showed that the data for group 1's responses to question one were not normalized. The mean score of group 1 was 3 while the mean score of group 2 was 3.52. The overall mean score of the two groups was 3.26. However there was no significant difference between the responses as shown by the two-sample t-test as the p-value was above .05.

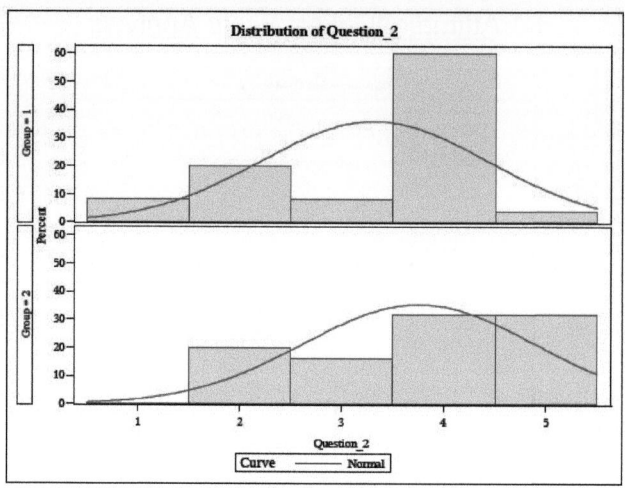

Figure 4.2 Distribution of Likert Scale Question 2

Question 2 of the MAEGUS study survey is "I know the scientific factors that determine how wind turbines generate energy." " with 1 meaning the participant strongly disagreed with the statement and 5 being the participant strongly agreed with the statement. The mean score of group 1 was 3.32 and the mean value of group 2 was 3.76. The overall mean score of the two groups was 3.54. There was no significant difference in the two groups results based on the t-test as the p-value was above .05.

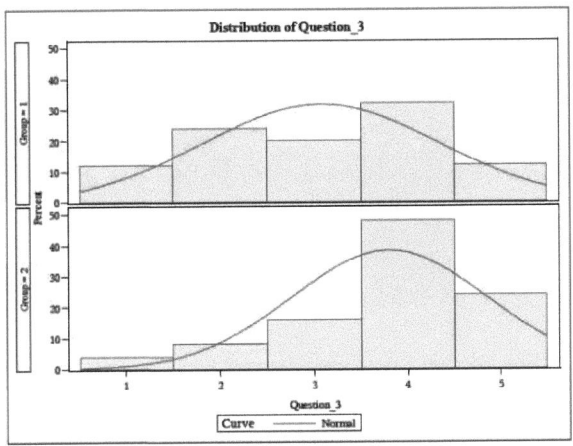

Figure 4.3 Distribution of Likert Scale Question 3

Question 3 of the MAEGUS survey is "I know the scientific factors that determine how solar arrays generate energy." " with 1 meaning the participant strongly disagreed with the statement and 5 being the participant strongly agreed with the statement. The mean score of group 1 was 3.08 while the mean value of group 2 was 3.8. The overall mean score of the two groups was 3.44. There was no significant difference in the two groups results based on the t-test as the p-value was above .05.

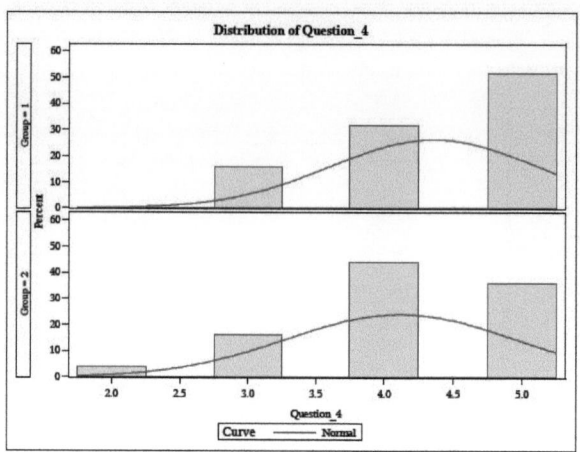

Figure 4.4. Distribution of Likert Scale Question 4

Question 4 of the MAEGUS survey is "I am interested in sustainable energy." with 1 meaning the participant strongly disagreed with the statement and 5 being the participant strongly agreed with the statement. The mean score of group 1 was 4.36 while group 2 was 4.12. The overall mean score of both groups was 4.24. There was no significant difference in the two groups results based on the t-test as the p-value was above .05.

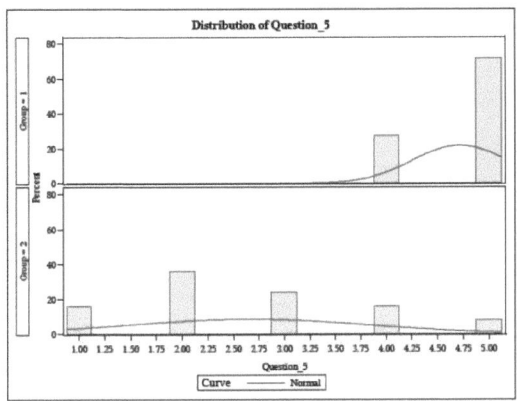

Figure 4.5. Distribution of Likert Scale Question 5

Question 5 of the MAEGUS survey is "I enjoyed the method I learned with today." with 1 meaning the participant strongly disagreed with the statement and 5 being the participant strongly agreed with the statement. The mean value of group 1 was 4.72 while the mean value of group 2 was 2.64. The overall mean score of the two groups was 3.68. There was a significant difference in the data between group 1 and group 2 as the p-value was less than .0001 and the confidence interval was positive. It is possible to reject the null hypothesis, that there is no difference in test results between a group given a serious game and a group with reading material, based on this result.

4.2 Cognitive Score Analysis

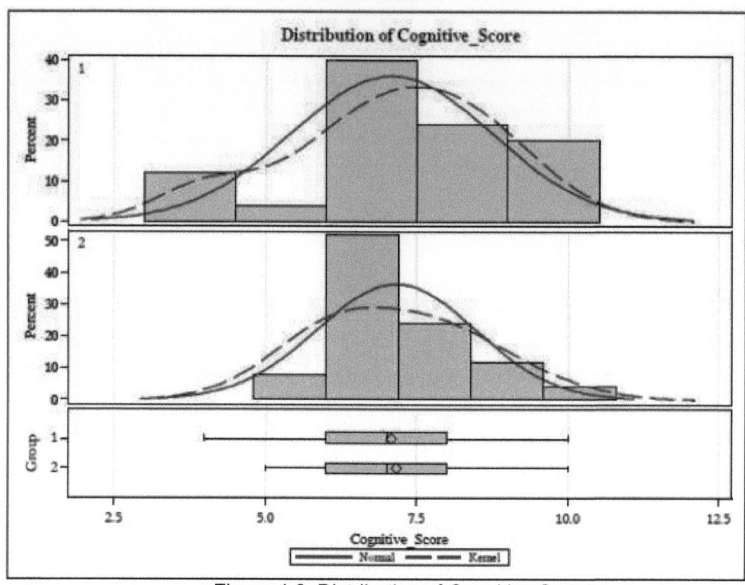

Figure 4.6. Distribution of Cognitive Scores

The cognitive scores were measured out of 10. Each question had only one correct answer and so each question had a weight of 1. The lowest score in group 1 was 4 while in group 2 the lowest score was 5. In both groups the highest score was a perfect score of 10. The mean score of group 1 was 7.08 and the mean score of group 2 was 7.16. The overall mean score of both groups was 7.12. There was no significant difference between the cognitive scores of both groups based on the t-test since the p-value was above 0.05. As such, the cognitive score cannot reject the null hypothesis.

The results presented here will be discussed in further detail and interpreted in the next section, the Conclusion.

CHAPTER 5. CONCLUSION

The MAEGUS study attempted to compare a serious game to more traditional reading material used in energy education curriculum to test its efficacy as a pedagogical application. Based on the results of the study, a number of interesting observations can be made.

5.1 Interpretation

Interpreting the results of the statistical analysis from the Results section, we find that MAEGUS, the serious game, is not significantly different than reading material from the Department of Energy in terms of test results. The cognitive scores on the tests were not significantly different and showed that participants in the study learned the same material from the game as they did the reading material. The attitudinal questions also didn't have any significant differences with the exception of which learning method participants liked better.

There is a standard distribution of answers to attitudinal question 5, the "I liked the method I learned with" question for group 2. The reading group had answers that varied from strongly negative to strongly positive enjoyment. However group 1 only had answers in the 4-5 range signifying that participants who played the game wholly had a positive experience. As expected participants in the MAEGUS study enjoyed playing the game more than participants who were placed in the reading group.

The null hypotheses of the MAEGUS study, There is no difference between test results of a group given a serious game versus a group given reading material, cannot be rejected based on the results of the study but that has significant value in itself. The conclusion of the MAEGUS study is that the serious game could be substituted for the 10 page reading material. Since there was no significant difference in the cognitive test scores, participants learned the same sustainable energy concepts playing the game as they would reading educational material. This shows that MAEGUS has some use as a pedagogical tool. More than that, it was the preferred method of learning and instilled the same level of confidence and interest in energy literacy as the reading material.

5.2 Discussion

There are some factors to take into consideration when comparing MAEGUS to educational material. In general participants took the full 20 minutes when playing the game and some participants still didn't have enough time to finish all 20 turns in game. However an overwhelming majority of participants in the reading group finished the information packet in the allotted 20 minutes. The researcher believes that the game takes longer to complete than the reading and so could not serve as a suitable substitute in terms of time.

A contributor to the longer time on task is that MAEGUS uses multiple data visualizations to provide information to the player about the energy generation formulas simulated in the game. Not all players were able to effectively identify and utilize the features of the data visualization and gave the visualizations a cursory glance or ignored them all together.

However, participants expressed interest in playing the game even if they were put in the reading group. One participant stated during the debriefing that they had a difficult time reading and was interested in a more visual method of learning.

Players were also highly motivated by the MAEGUS high score system. A number of players stated that they wanted to learn how the formula worked so they could get the high score. One participant asked for scratch paper so he could calculate how the upgrades would affect his energy generation.

Players were excited to find out if they got a high score at the end and liked seeing how a variety of strategies created the high scores. Many participants stated that they were interested in playing the game again especially since they felt they understood the game better after the first play through.

Finally, the results seem to favor MAEGUS being used to help teach students who have difficulty engaging with reading material. As such MAEGUS may have the best use with a younger K-12 audience, especially high school students, and can increase awareness and interest in sustainable energy paths in secondary education.

5.3 Future Work

In terms of future research, the MAEGUS serious game could use further validation. The MAEGUS study survey was created based off of DeWaters' Energy Literacy Survey (DeWaters, 2008). However the Energy Literacy Survey was developed for high school level students and the MAEGUS study survey was used on college students who may have a better understanding of sustainable energy from their college courses. In future research a more rigorous scale for energy literacy should be developed for MAEGUS. The MAEGUS study was a serious game study as opposed to research into developing an educational questionnaire and so a higher focus was placed on development and evaluation of the MAEGUS game.

In order to test true retention of energy concepts, a longitudinal study with MAEGUS should be conducted similar to Darling's Racing Academy study which took place over a number of weeks as opposed to one single gameplay session (Darling, 2008). In the longitudinal study, players would take a pre-test on their conceptual knowledge of sustainable energy and then begin a learning module where they learned about sustainable energy concepts using MAEGUS gameplay sessions and other teaching methods. A post-test would be administered after the learning module was concluded and the results of the post-test would be compared to results from the pre-test. Further, the post-test could be compared to a group that underwent the same learning module with the exclusion of MAEGUS gameplay sessions.

As a serious game, MAEGUS also has room for further development. One future improvement for the game would be adding more renewable resource simulations such as hydropower, geothermal and even nuclear energy sources. Even nonrenewable resources such as coal and oil could develop a baseline for comparing sustainable versus non-sustainable energy sources. Participants also pointed out that the upgrade system didn't have a substantial effect on their energy generation to utilize. Upgrades, like all of the other factors of energy generation, were realistically simulated and participants discovered that building new energy generators as opposed to upgrading was more cost effective and produced more energy. This gameplay issue was only discovered during playtesting which is an indicator that MAEGUS needs further game balancing.

Improvement of the GUI, more levels, expanded scenarios and further framework development in the future could improve MAEGUS's ability to engage and teach players.

5.4 Final Thoughts

In conclusion, this paper discusses the development and test of the MAEGUS serious game, a new pedagogical application that can help teach sustainable energy concepts. The results of the MAEGUS testing showed that MAEGUS is an effective and engaging tool and could have a role supplementing normal energy education curriculum. While it is not the ultimate substitute for traditional teaching methods, such as reading material, it is an effective and noteworthy option for improving energy literacy in a novel and fun manner.

LIST OF REFERENCES

LIST OF REFERENCES

Acikgoz, C. (2011). Renewable energy education in Turkey. *Renewable Energy, 36*(2), 608-611.

Belu, R., & Koracin, D. (2012, May). Effects of complex wind regimes and meteorlogical parameters on wind turbine performances. In *Energytech, 2012 IEEE* (pp. 1-6). IEEE.

Chen, K. L., Huang, S. H., & Liu, S. Y. (2013). Devising a framework for energy education in Taiwan using the analytic hierarchy process. *Energy Policy, 55*(1), 396-403.

Chen, X., Wang, F. J., Liu, T. Q., Chen, Z. H., Li, X. H., & Guan, T. Y. (2012, March). Wind Power Prediction Considering the Layout of the Wind Turbines and Wind Direction. In *Power and Energy Engineering Conference (APPEEC), 2012 Asia-Pacific* (pp. 1-4). IEEE.

Csikszentmihalyi, M. (1975). *Beyond Boredom and Anxiety*. San Francisco: Josey-Bass.

Darling, J., Drew, B., Joiner, R., Iacovides, I., & Gavin, C. (2008). Game Based Learning in Engineering Education. In *International Conference on Innovation, Good Practice and Research in Engineering Education, Loughborough University, England P* (Vol. 70, p. 2008).

de Freitas, S., Kiili, K., Ney, M., Ott, M., Popescu, M., Romero, M., & Stanescu, I. (2012). GEL: Exploring Game Enhanced Learning. *Procedia Computer Science, 15*, 289-292.

DeWaters, J., & Powers, S. (2008). Energy literacy among middle and high school youth. *2008 38th Annual Frontiers in Education Conference*, T2F–6–T2F–11.

DeWaters, J. E., & Powers, S. E. (2011). Energy literacy of secondary students in New York State (USA): A measure of knowledge, affect, and behavior. *Energy Policy, 39*(3), 1699-1710.

Dickey, O. M. D. (2006). Game Design Narrative for Learning : Appropriating Adventure Game Design Narrative Devices and Techniques for the Design of Interactive Learning Environments. *Educational Technology Research and Development, 54*(3), 245–263.

Foley, J. T., & Gutowski, T. G. (2008). TurbSim: Reliability-based wind turbine simulator. *2008 IEEE International Symposium on Electronics and the Environment*, 1–5.

Garg, H. P., & Kandpal, T. C. (1996). Renewable energy education: challenges and problems in developing countries. *Renewable energy, 9*(1), 1188-1193.

Green, M. A., Emery, K., Hishikawa, Y., Warta, W., Dunlop, E.D. (2013) Solar cell efficiency tables (version 41). *Progress in Photovoltaics, 21*(1), 1-11.

Jennings, P., & Lund, C. (2001). Renewable energy education for sustainable development. *Renewable Energy, 22*(1), 113

Kandpal, T. C., & Garg, H. P. (1999). Energy education. *Applied Energy, 64*(1), 71-78.

Kaygusuz, K. (2010). Wind energy status in renewable electrical energy production in Turkey. *Renewable and Sustainable Energy Reviews, 14*(7), 2104-2112.

Kaygusuz, K. (2012). Energy for sustainable development: A case of developing countries. *Renewable and Sustainable Energy Reviews, 16*(2), 1116-1126.

Kiili, K., de Freitas, S., Arnab, S., & Lainema, T. (2012). The Design Principles for Flow Experience in Educational Games. *Procedia Computer Science*, 15, 78–91.

Koussa, M., Cheknane, A., Hadji, S., Haddadi, M., & Noureddine, S. (2011). Measured and modeled improvement in solar energy yield from flat plate photovoltaic systems utilizing different tracking systems and under a range of environmental conditions. *Applied Energy*, *88*(5), 1756-1771.

Mayer, I. (2012). Towards a Comprehensive Methodology for the Research and Evaluation of Serious Games. *Procedia Computer Science*, 15, 233–247.

Mehleri, E. D., Zervas, P. L., Sarimveis, H., Palyvos, J. A., & Markatos, N. C. (2010). Determination of the optimal tilt angle and orientation for solar photovoltaic arrays. *Renewable energy*, *35*(11), 2468-2475.

Nataraja, K., Head, N. & Whittinghill, D. (2014). An empirical test of an educational video game to teach energy literacy. *ASEE Illinois-Indiana Section Conference, Terre Haute, IN.*

"National Renewable Energy Laboratory (NREL) Home Page." *National Renewable Energy Laboratory (NREL) Home Page*. Web. 21 Feb. 2014.

Saidur, R., Islam, M. R., Rahim, N. A., & Solangi, K. H. (2010). A review on global wind energy policy. *Renewable and Sustainable Energy Reviews*, *14*(7), 1744-1762.

Stone, C. (2011, October). Renewable energy education at the Colorado School of Mines: A survey of development. In *Frontiers in Education Conference (FIE), 2011* (pp. S2H-1). IEEE.

Wilks, D. S., & Wilby, R. L. (1999). The weather generation game: a review of stochastic weather models. *Progress in Physical Geography*, 23(3), 329–357.

Wu, W.-H., Chiou, W.-B., Kao, H.-Y., Alex Hu, C.-H., & Huang, S.-H. (2012). Re-exploring game-assisted learning research: The perspective of learning theoretical bases. *Computers & Education*, 59(4), 1153–1161.

Yue, C. D., & Huang, G. R. (2011). An evaluation of domestic solar energy potential in Taiwan incorporating land use analysis. *Energy Policy*, *39*(12), 7988-8002.

APPENDICES

Appendix A. MAEGUS Study

Subject No: _____
Group: _____
Major: _____
Classification: _____

Please answer the following questions using the provided scale, 1 signifying you do not agree
and 5 signifying you definitely agree.

1. I have an in-depth knowledge of sustainable energy.

1	2	3	4	5
Strongly Disagree	Slightly Disagree	Neutral	Slightly Agree	Strongly Agree

2. I know the scientific factors that determine how wind turbines generate energy.

1	2	3	4	5
Strongly Disagree	Slightly Disagree	Neutral	Slightly Agree	Strongly Agree

3. I know the scientific factors that determine how solar arrays generate energy.

1	2	3	4	5
Strongly Disagree	Slightly Disagree	Neutral	Slightly Agree	Strongly Agree

4. I am interested in sustainable energy.

1	2	3	4	5
Strongly Disagree	Slightly Disagree	Neutral	Slightly Agree	Strongly Agree

5. I enjoyed the method I learned with today.

1	2	3	4	5
Strongly Disagree	Slightly Disagree	Neutral	Slightly Agree	Strongly Agree

Please answer the following multiple choice questions to the best of your ability.

1. The amount of ELECTRICAL ENERGY (ELECTRICITY) we use is measured in units called...
A. Kilowatts (kW)
B. Kilowatt-hours (kW-h)
C. British Thermal Units (BTU)
D. Volts (V)
E. Horsepower (HP)

2. What does it mean if an electric power plant is 35% efficient?
A. For every $100 invested in the production of energy, $35 is made into profit
B. For every $35 invested in the production of energy, $100 is made into profit
C. For every 100 units of energy that go into the plant, 35 units are lost during energy transformations
D. For every 100 units of energy that go into the plant, 35 units are converted into electrical
energy
E. For every 35 units of energy that go into the plant, 100 units of electrical energy are produced

3. The term "renewable energy resources" means ...
A. Resources that are free and convenient to use
B. Resources that can be converted directly into heat and electricity
C. Resources that do not produce air pollution
D. Resources that are very efficient to use for producing energy
E. Resources that can be replenished by nature in a short period of time

4. Which of the following energy resources is **NOT** renewable?
A. Solar
B. Biomass (wood, waste, plants, alcohol fuels)
C. Coal
D. Water (hydro) power
E. Geothermal

5. Solar electricity generators absorb energy from the sun and are also called...
A. Sun Trackers
B. Solar Absorbers
C. Photon Vehicular Trackers
D. Photovoltaic Trackers
E. Solar Converters

6. Wind Turbines derive energy by...
A. Converting mechanical energy when turned by the wind
B. Absorbing heat in the air as they turn
C. Perpetually turning and generating mechanical energy
D. Absorbing moisture in the air
E. None of the above

7. As a wind turbine's height increases the following occurs...
A. Energy generation increases because air density decreases.
B. Energy generation increases because wind speed increases.
C. Energy generation decreases because air density increases.
D. Energy generation decreases because wind speed decreases.
E. None of the above

8. On average, how much energy (kwh) does it take to power a household?
A. 3 kwh
B. 10 kwh
C. 30 kwh
D. 100 kwh
E. 300 kwh

9. Which of the following is not a factor of wind energy generation?
A. Wind Velocity
B. Swept area of the blades
C. Air Density
D. Air Temperature
E. Generator efficiency

10. Which of the following is not a factor of solar energy generation?
A. Solar Concentration
B. Estimated Wattage
C. Internal Battery Life
D. Average Sun Hours
E. All of the above are factors of solar energy generation

Appendix B. Reading Material

ENERGY
EFFICIENCY
AND
RENEWABLE
ENERGY

CLEARINGHOUSE

DOE/GO-102001-1102
FS175
March 2001

Renewable Energy: An Overview

What is Renewable Energy?

Renewable energy uses energy sources that are continually replenished by nature—the sun, the wind, water, the Earth's heat, and plants. Renewable energy technologies turn these fuels into usable forms of energy—most often electricity, but also heat, chemicals, or mechanical power.

Why Use Renewable Energy?

Today we primarily use fossil fuels to heat and power our homes and fuel our cars. It's convenient to use coal, oil, and natural gas for meeting our energy needs, but we have a limited supply of these fuels on the Earth. We're using them much more rapidly than they are being created. Eventually, they will run out. And because of

safety concerns and waste disposal problems, the United States will retire much of its nuclear capacity by 2020. In the meantime, the nation's energy needs are expected to grow by 33 percent during the next 20 years. Renewable energy can help fill the gap.

Even if we had an unlimited supply of fossil fuels, using renewable energy is better for the environment. We often call renewable energy technologies "clean" or "green" because they produce few if any pollutants. Burning fossil fuels, however, sends greenhouse gases into the atmosphere, trapping the sun's heat and contributing to global warming. Climate scientists generally agree that the Earth's average temperature has risen in the past century. If this trend continues, sea levels will rise, and scientists predict that floods, heat waves, droughts, and other extreme weather conditions could occur more often.

Other pollutants are released into the air, soil, and water when fossil fuels are burned. These pollutants take a dramatic toll on the environment—and on humans. Air pollution contributes to diseases like asthma. Acid rain from sulfur dioxide and nitrogen oxides harms plants and fish. Nitrogen oxides also contribute to smog.

A PV system at the Pinnacles National Monument in California eliminates a $20,000 annual fuel bill for a diesel generator that produced each year 149 tons of carbon dioxide—a greenhouse gas.

This document was produced for the U.S. Department of Energy (DOE) by the National Renewable Energy Laboratory (NREL), a DOE national laboratory. The document was produced by the Information and Outreach Program at NREL for the DOE Office of Energy Efficiency and Renewable Energy. The Energy Efficiency and Renewable Energy Clearinghouse (EREC) is operated by NCI Information Systems, Inc., for NREL/DOE. The statements contained herein are based on information known to EREC and NREL at the time of printing. No recommendation or endorsement of any product or service is implied if mentioned by EREC.

geothermal energy. Direct-use applications require geothermal temperatures between about 70° to 302°F—lower than those required for electricity generation. The United States already has about 1,300 geothermal direct-use systems in operation.

In a direct-use system, a well is drilled into a geothermal reservoir, which provides a steady stream of hot water. Some systems use the water directly, but most pump the water through what's called a *heat exchanger*. The heat exchanger keeps the water separate from a working fluid (usually water or a mixture of water and antifreeze), which is heated by the geothermal water. The working fluid then flows through piping, distributing the heat directly for its intended use.

The heated water or fluid can be used in a building to replace the traditional heat source—often natural gas—of a boiler, furnace, and hot water heater. Some cities and towns actually have large direct-use heating systems—called district heating—that provide many buildings with heat. Geothermal direct use is also used in agriculture—such as for fish farms and to heat greenhouses—and for industrial food processing (vegetable dehydration).

Geothermal Heat Pumps

While air temperatures can vary widely through the seasons, the temperatures of the shallow ground only range from 50° to 70°F depending on latitude. GHPs draw on this relatively stable temperature as a source for heating buildings in the winter and keeping them cool in the summer.

Through underground piping, a GHP discharges heat from inside a building into the ground in the summer, much like a refrigerator uses electricity to keep its interior cool while releasing heat into your kitchen. In the winter, this process is reversed; the GHP extracts heat from the ground and releases it into a building.

Because GHPs actually move heat between homes and the earth, instead of burning fuels, they operate very cleanly and efficiently. In fact, GHPs are at least three times more efficient than even the most energy-efficient furnaces on the market today.

Solar Energy

Solar technologies tap directly into the infinite power of the sun and use that energy to produce heat, light, and power.

Passive Solar Lighting and Heating

People have used the sun to heat and light their homes for centuries. Ancient Native Americans built their dwellings directly into south-facing cliff walls because they knew the sun travels low across the southern sky in the Northern Hemisphere during the winter. They also knew the massive rock of the cliff would absorb heat in winter and protect against wind and snow. At the same time, the cliff-dwelling design blocked sunlight during the summer, when the sun is higher in the sky, keeping their dwellings cool.

The modern version of this sun-welcoming design is called *passive solar* because no pumps, fans, or other mechanical devices are used. Its most basic features include large, south-facing windows that fill the home with natural sunlight, and dark tile or brick floors that store the sun's heat and release it back into the home at night. In the summer, when the sun is higher in the sky, window overhangs block direct sunlight, which keeps the house cool. Tile and brick floors also remain cool during the summer.

Passive solar design combined with energy efficiency will go even further. Energy-efficient features such as energy-saving windows and appliances, along with good insulation and weatherstripping, can make a huge difference in energy and cost savings.

Solar Water Heating

Solar energy can be used to heat water for your home or your swimming pool. Most solar water-heating systems consist of a solar collector and a water storage tank.

Solar water-heating systems use collectors, generally mounted on a south-facing roof, to heat either water or a heat-transfer fluid, such as a nontoxic antifreeze. The heated water is then stored in a water tank similar to one used in a conventional gas or electric water-heating system.

Altogether about 2800 MW of geothermal electric capacity is produced annually in this country.

This homeowner in Aurora, Colorado, uses a GHP to heat and cool his home.

4

The Four Times Square Building in New York City uses thin-film PV panels to reduce the building's power load from the utility grid.

Passive solar building techniques turn homes into huge solar collectors.

There are basically three types of solar collectors for heating water: flat-plate, evacuated-tube, and concentrating. The most common type, a *flat-plate collector,* is an insulated, weatherproof box containing a dark absorber plate under a transparent cover. *Evacuated-tube collectors* are made up of rows of parallel, transparent glass tubes. Each tube consists of a glass outer tube and an inner tube, or absorber, covered with a coating that absorbs solar energy but inhibits heat loss. *Concentrating collectors* for residential applications are usually parabolic-shaped mirrors (like a trough) that concentrate the sun's energy on an absorber tube called a receiver that runs along the axis of the mirrored trough and contains a heat-transfer fluid.

All three types of collectors heat water by circulating household water or a heat-transfer fluid such as a nontoxic antifreeze from the collector to the water storage tanks. Collectors do this either passively or actively.

Passive solar water-heating systems use natural convection or household water pressure to circulate water through a solar collector to a storage tank. They have no electric components that could break, a feature that generally makes them more reliable, easier to maintain, and possibly longer lasting than active systems.

An *active* system uses an electric pump to circulate water or nontoxic antifreeze through the system. Active systems are usually more expensive than passive systems, but they are also more efficient. Active systems also can be easier to retrofit than passive systems because their storage tanks do not need to be installed above or close to the collectors. Also, the moving water in the system will not freeze in cold climates. But because these systems use electricity, they will not function in a power outage. That's why many active systems are now combined with a small solar-electric panel to power the pump.

The amount of hot water a solar water heater produces depends on the type and size of the system, the amount of sun available at the site, proper installation, and the tilt angle and orientation of the collectors. But if you're currently using an electric water heater, solar water heating is a cost-effective alternative. If you own a swimming pool, heating the water with solar collectors can also save you money.

Solar Electricity

Solar electricity or photovoltaic (PV) technology converts sunlight directly into electricity. Solar electricity has been a prime source of power for space vehicles since the inception of the space program. It has also been used to power small electronics and rural and agricultural applications for three decades. During the last decade, a strong solar electric market has emerged for powering urban grid-connected homes and buildings as a result of advances in solar technology along with global changes in electric industry restructuring.

Although many types of solar electric systems are available today, they all consist of basically three main items: *modules* that convert sunlight into electricity; *inverters* that convert that electricity into alternating current so it can be used by most household appliances; and possibly or sometimes *batteries* that store excess electricity produced by the system. The remainder of the system comprises equipment such as wiring, circuit breakers, and support structures.

Today's modules can be built into glass skylights and walls. Some modules resemble traditional roof shingles, but they generate electricity, and some come with built-in inverters. The solar modules available today are more efficient and versatile than ever before.

In over 30 states, any additional power produced by a PV system, which is not being used by a home or building, can be fed back to the electric grid through a process known as *net metering*. Net metering allows electricity customers to pay only for their "net" electricity, or the

5

amount of power consumed from their utility minus the power generated by their PV system. This metering arrangement allows consumers to realize full retail value for 100 percent of the PV energy produced by their systems.

Grid-connected PV systems do not require batteries. However, some grid connected systems use them for emergency backup power. And of course in remote areas, solar electricity is often a economic alternative to expensive distribution line extensions incurred by a customer first connecting to the utility grid. Electricity produced by solar electric systems in remote locations is stored in batteries. Batteries will usually store electricity produced by a solar-electric system for up to three days.

This dish/Stirling solar power system in Arizona is capable of producing 25 kW of electricity.

What type of system to purchase will depend on the energy-efficiency of your home, your home's location, and your budget. Before you size your system, try reducing energy demand through energy-efficient measures. Purchasing energy-saving appliances and lights, for example, will reduce your electrical demand and allow you to purchase a smaller solar-electric system to meet your energy needs or get more value from a larger system. Energy efficiency allows you to start small and then add on as your energy needs increase.

Solar Thermal Electricity
Unlike solar-electric systems that convert sunlight into electricity, solar thermal electric systems convert the sun's heat into electricity. This technology is used primarily in large-scale power plants for powering cities and communities, especially in the Southwest where consistent hours of sunlight are greater than other parts of the United States.

Concentrating solar power (CSP) technologies convert solar energy into electricity by using mirrors to focus sunlight onto a component called a receiver. The receiver transfers the heat to a conventional engine-generator—such as a steam turbine—that generates electricity.

There are three types of CSP systems: power towers (central receivers), parabolic troughs, and dish/engine systems. A *power tower system* uses a large field of

mirrors to concentrate sunlight onto the top of a tower, where a receiver sits. Molten salt flowing through the receiver is heated by the concentrated sunlight. The salt's heat is turned into electricity by a conventional steam generator. *Parabolic-trough systems* concentrate the sun's energy through long, parabolic-shaped mirrors. Sunlight is focused on a pipe filled with oil that runs down the axis of the trough. When the oil gets hot, it is used to boil water in a conventional steam generator to produce electricity. A *dish/engine system* uses a mirrored dish (similar in size to a large satellite dish). The dish-shaped surface focuses and concentrates the sun's heat onto a receiver at the focal point of the dish (above and center of the collectors). The receiver absorbs the sun's heat and transfers it to a fluid within an engine, where the heat causes the fluid to expand against a piston to produce mechanical power. The mechanical power is then used to run a generator or alternator to produce electricity.

Concentrating solar technologies can be used to generate electricity for a variety of applications, ranging from remote power systems as small as a few kilowatts (kW) up to grid-connected applications of 200 MW or more. A 354-MW power plant in Southern California, which consists of nine trough power plants, meets the energy needs of more than 350,000 people and is the world's largest solar energy power plant.

Wind Energy

For hundreds of years, people have used windmills to harness the wind's energy. Today's wind turbines, which operate differently from windmills, are a much more efficient technology.

Wind turbine technology may look simple: the wind spins turbine blades around a central hub; the hub is connected to a shaft, which powers a generator to make electricity. However, turbines are highly sophisticated power systems that capture the wind's energy by means of new blade designs or *airfoils*. Modern mechanical drive systems, combined with advanced generators, convert that energy into electricity.

Wind turbines that provide electricity to the utility grid range in size from 50 kW to

Wind energy has been the fastest growing source of energy since 1990...

6

53

The 6-MW Green Mountain power plant in Searsburg, Vermont, consists of eleven 550-kW wind turbines.

Hydrogen is high in energy, yet its use as a fuel produces water as the only emission.

NASA uses liquid hydrogen to launch its space shuttles and hydrogen fuel cells to provide them with electricity

1 or 2 MW. Large, utility-scale projects can have hundreds of turbines spread over many acres of land. Small turbines, below 50 kW, are used to charge batteries, electrify homes, pump water for farms and ranches, and power remote telecommunications equipment. Wind turbines can also be placed in the shallow water near a coastline if open land is limited, such as in Europe, and/or to take advantage of strong, offshore winds.

Wind energy has been the fastest growing source of energy in the world since 1990, increasing at an average rate of over 25 percent per year. It's a trend driven largely by dramatic improvements in wind technology. Currently, wind energy capacity amounts to about 2500 MW in the United States. Good wind areas, which cover 6 percent of the contiguous U.S. land area, could supply more than one and a half times the 1993 electricity consumption of the entire country.

California now has the largest number of installed turbines. Many turbines are also being installed across the Great Plains, reaching from Montana east to Minnesota and south through Texas, to take advantage of its vast wind resource. North Dakota alone has enough wind to supply 36 percent of the total 1990 electricity consumption of the lower 48 states. Hawaii, Iowa, Minnesota, Oregon, Texas, Washington, Wisconsin, and Wyoming are among states where wind energy use is rapidly increasing.

Hydrogen

Hydrogen is high in energy, yet its use as a fuel produces water as the only emission. Hydrogen is the universe's most abundant element and also its simplest. A hydrogen atom consists of only one proton and one electron. Despite its abundance and simplicity, it doesn't occur naturally as a gas on the Earth.

Today, industry produces more than 4 trillion cubic feet of hydrogen annually. Most of this hydrogen is produced through a process called *reforming*, which involves the application of heat to separate hydrogen from carbon. Researchers are developing highly efficient, advanced reformers to produce hydrogen from natural gas for what's called *Proton Exchange Membrane* fuel cells.

You can think of fuel cells as batteries that never lose their charge. Today, hydrogen fuel cells offer tremendous potential to produce electrical power for distributed energy systems and vehicles. In the future, hydrogen could join electricity as an important "energy carrier": storing, moving, and delivering energy in a usable form to consumers. Renewable energy sources, like the sun, can't produce energy all the time. But hydrogen can store the renewable energy produced until it's needed.

Eventually, researchers would like to directly produce hydrogen from water using solar, wind, and biomass and biological technologies.

Ocean Energy

The ocean can produce two types of energy: *thermal energy* from the sun's heat, and *mechanical energy* from the tides and waves.

Ocean thermal energy can be used for many applications, including electricity generation. Electricity conversion systems use either the warm surface water or boil the seawater to turn a turbine, which activates a generator.

The electricity conversion of both tidal and wave energy usually involves mechanical devices. A dam is typically used to convert tidal energy into electricity by forcing the water through turbines, activating a generator. Meanwhile, wave energy uses mechanical power to directly activate a generator, or to transfer to a working fluid, water, or air, which then drives a turbine/generator.

Most of the research and development in ocean energy is happening in Europe.

7

2.2 TODAY'S COMMERCIAL WIND TECHNOLOGY

Beginning with the birth of modern wind-driven electricity generators in the late 1970s, wind energy technology has improved dramatically up to the present. Capital costs have decreased, efficiency has increased, and reliability has improved. High-quality products are now routinely delivered by major suppliers of turbines around the world, and complete wind generation plants are being engineered into the grid infrastructure to meet utility needs. In the 20% Wind Scenario outlined in this report, it is assumed that capital costs would be reduced by 10% over the next two decades, and capacity factors would be increased by about 15% (corresponding to a 15% increase in annual energy generation by a wind plant).

2.2.1 WIND RESOURCES

Wind technology is driven by the nature of the resource to be harvested. The United States, particularly the Midwestern region from Texas to North Dakota, is rich in wind energy resources as shown in Figure 2-1, which illustrates the wind resources measured at a 50-meter (m) elevation. Measuring potential wind energy generation at a 100-m elevation (the projected operating hub height of the next generation of modern turbines) greatly increases the U.S. land area that could be used for wind deployment, as shown in Figure 2-2 for the state of Indiana. Taking these measurements into account, current U.S. land-based and offshore wind resources are estimated to be sufficient to supply the electrical energy needs of the entire country several times over. For a description of U.S. wind resources, see Appendix B.

Figure 2-1. The wind resource potential at 50 m above ground on land and offshore

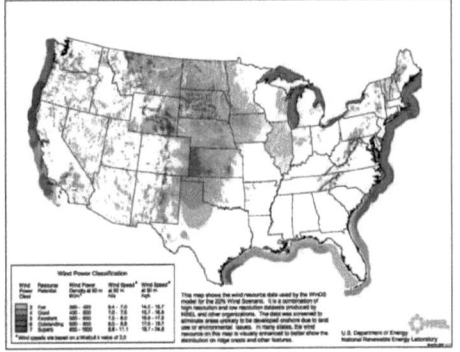

Identifying the good wind potential at high elevations in states such as Indiana and off the shore of both coasts is important because it drives developers to find ways to harvest this energy. Many of the opportunities being pursued through advanced

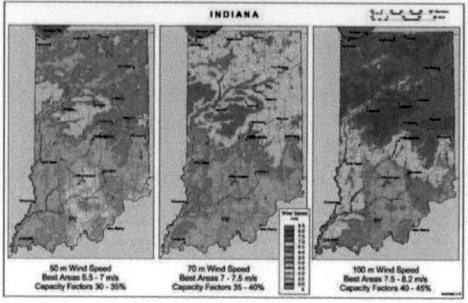

technology are intended to achieve higher elevations, where the resource is much
greater, or to access extensive offshore wind resources.

2.2.2 TODAY'S MODERN WIND TURBINE

Modern wind turbines, which are currently being deployed around the world, have
three-bladed rotors with diameters of 70 m to 80 m mounted atop 60-m to 80-m
towers, as illustrated in Figure 2-3. Typically installed in arrays of 30 to 150
machines, the average turbine installed in the United States in 2006 can produce
approximately 1.6 megawatts (MW) of electrical power. Turbine power output is
controlled by rotating the blades around their long axis to change the angle of attack
with respect to the relative wind as the blades spin around the rotor hub. This is
called controlling the blade pitch. The turbine is pointed into the wind by rotating
the nacelle around the tower. This is called controlling the yaw. Wind sensors on the
nacelle tell the yaw controller where to point the turbine. These wind sensors, along
with sensors on the generator and drivetrain, also tell the blade pitch controller how
to regulate the power output and rotor speed to prevent overloading the structural
components. Generally, a turbine will start producing power in winds of about
5.36 m/s and reach maximum power output at about 12.52 m/s–13.41 m/s. The
turbine will pitch or feather the blades to stop power production and rotation at
about 22.35 m/s. Most utility-scale turbines are upwind machines, meaning that they
operate with the blades upwind of the tower to avoid the blockage created by the
tower.

The amount of energy in the wind available for extraction by the turbine increases
with the cube (the third power) of wind speed; thus, a 10% increase in wind speed
creates a 33% increase in available energy. A turbine can capture only a portion of
this cubic increase in energy, though, because power above the level for which the
electrical system has been designed, referred to as the rated power, is allowed to
pass through the rotor.

56

Figure 2-3. A modern 1.5-MW wind turbine installed in a wind power plant

Rotor Hub

Tower, 80 m

Minivan

Rotor Blades:
* Shown Feathered
* Length, 37 m

Nacelle Enclosing:
* Low-Speed Shaft
* Gearbox
* Generator, 1.5 MW
* Electrical Controls

In general, the speed of the wind increases with the height above the ground, which is why engineers have found ways to increase the height and the size of wind turbines while minimizing the costs of materials. But land-based turbine size is not expected to grow as dramatically in the future as it has in the past. Larger sizes are physically possible; however, the logistical constraints of transporting the components via highways and of obtaining cranes large enough to lift the components present a major economic barrier that is difficult to overcome. Many turbine designers do not expect the rotors of land-based turbines to become much larger than about 100 m in diameter, with corresponding power outputs of about 3 MW to 5 MW.

2.2.3 WIND PLANT PERFORMANCE AND PRICE

The performance of commercial turbines has improved over time, and as a result, their capacity factors have slowly increased. Figure 2-4 shows the capacity factors at commercial operation dates (CODs) ranging from 1998 to 2005. The data show that turbines in the Lawrence Berkeley National Laboratory (Berkeley Lab) database (Wiser and Bolinger 2007) that began operating commercially before 1998 have an average capacity factor of about 22%. The turbines that began commercial operation after 1998, however, show an increasing capacity factor trend, reaching 36% in 2004 and 2005.

The cost of wind-generated electricity has dropped dramatically since 1980, when the first commercial wind plants began operating in California. Since 2003, however, wind energy prices have increased. Figure 2-5 (Wiser and Bolinger 2007)

Background

What is a solar electric or photovoltaic system?

Photovoltaic (PV) systems convert sunlight directly to electricity. They work any time the sun is shining, but more electricity is produced when the sunlight is more intense and strikes the PV modules directly (as when rays of sunlight are perpendicular to the PV modules). Unlike solar thermal systems for heating water, PV does not use the sun's heat to make electricity. Instead, electrons freed by the interaction of sunlight with semiconductor materials in PV cells are captured in an electric current.

PV allows you to produce electricity—without noise or air pollution—from a clean, renewable resource. A PV system never runs out of fuel, and it won't increase U.S. oil imports. Many PV system components are manufactured right here in the United States. These characteristics could make PV technology the U.S. energy source of choice for the 21st century.

The basic building block of PV technology is the solar "cell." Multiple PV cells are connected to form a PV "module," the smallest PV component sold commercially. Modules range in power output from about 10 watts to 300 watts. A PV system connected or "tied" to the utility grid has these components:

• One or more PV modules, which are connected to an inverter

• The inverter, which converts the system's direct-current (DC) electricity to alternating current (AC)

• Batteries (optional) to provide energy storage or backup power in case of a power interruption or outage on the grid.

AC electricity is compatible with the utility grid. It powers our lights, appliances, computers, and televisions.

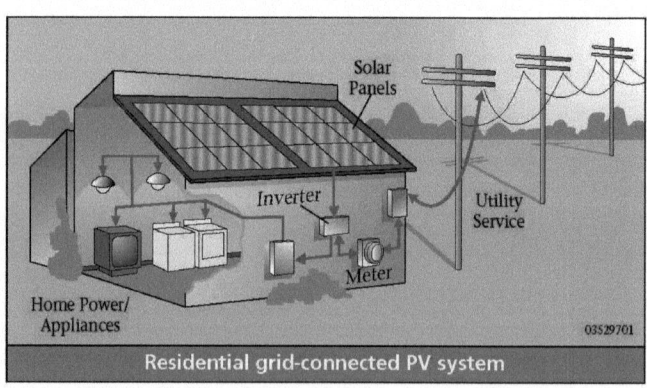

Solar Panels

Inverter

Utility Service

Meter

Home Power/ Appliances

03529701

Residential grid-connected PV system

2

Investing in a PV system

Why should you buy a PV system?

People decide to buy PV systems for a variety of reasons. Some people want to help preserve the Earth's finite fossil-fuel resources and reduce air pollution. Others want to invest in an energy-producing improvement to their property. Some people like the security of reducing the amount of electricity they buy from their utility because it makes them less vulnerable to future price increases. And some people just appreciate the independence that a PV system provides.

If you plan to build a home away from an established utility service, inquire about the cost of installing a utility line. Often, the cost of extending conventional power to your residence is higher than the cost of a solar option.

Whatever your reason, solar energy is widely thought to be the energy source of choice for the future, and you may be able to take advantage of a state-sponsored program to help make it your energy choice for today and tomorrow.

Is your home or business a good place for a PV system?

Can you locate your system so it works well?

A well-designed PV system needs clear and unobstructed access to the sun's rays for most or all of the day, throughout the year. You can make an initial assessment yourself. If the location looks promising, your PV provider can determine whether your home or business can effectively use a PV system.

The orientation of your PV system (the compass direction that your system faces) affects its performance. In the United States, the sun is always in the southern half of the sky but is higher in the summer and lower in the winter. Usually, the best location for a PV system is a south-facing roof, but roofs that face east or west may also be acceptable. Flat roofs also work well for solar electric systems, because PV modules can be mounted flat on the roof facing the sky or bolted on frames tilted toward the south at an optimal angle. They can also be attached directly to the roof as "PV shingles."

If a rooftop can't be used, your solar modules can also be placed on the ground, either on a fixed mount or a "tracking" mount that follows the sun to orient the PV modules. Other options (often used in multifamily or commercial applications) include mounting structures that create covered parking, or that provide shade as window awnings.

Is your site free from shading by trees, nearby buildings, or other obstructions?

To make the best use of your PV system, the PV modules must have a clear "view" of the sun for most or all of the day—unobstructed by trees, roof gables, chimneys, buildings, and other features of your home and the

5

59

surrounding landscape. Some potential sites for your PV system may be bright and sunny during certain times of the day, but shaded during other times. Such shading may substantially reduce the amount of electricity that your system will produce. To be eligible for some rebates, your system must be unshaded between certain hours during certain times of the year. Some states have laws that establish your right to protect your solar access through the creation of a "solar easement." Your PV provider can help you determine whether your site is suitable for a solar electric system.

Does your roof or property contain a large enough area for the PV system?

The amount of space that a PV system needs depends on the size of the system you purchase. Some residential systems require as little as 50 square feet (for a small "starter" system), but others could need as much as 1,000 square feet. Commercial systems are typically even larger. If your location limits the size of your system, you may want to install one that uses more efficient PV modules. Greater efficiency means that the module needs less surface area to convert sunlight into a given amount of electric power. PV modules are available in a range of types, and some offer more efficiency per square foot than others do (see table on the next page). Although the efficiency (percent of sunlight converted to electricity) varies with the different types of PV modules available today, higher efficiency modules typically

cost more. System sizing, discussed later in this booklet, should also be discussed with your PV provider.

What kind of roof do you have, and what is its condition?

Some types of roofs are simpler and cheaper to work with, but a PV system can be installed on any type. Typically, roofs with composition shingles are the easiest to work with, and those with slate are the most difficult. In any case, an experienced solar installer will know how to work on all types and can use roofing techniques that eliminate any possibility of leaks. Ask your PV provider how the PV system affects your roof warranty.

If your roof is older and needs to be replaced in the near future, you may want to replace it at the time the PV system is installed to avoid the cost of removing and reinstalling your PV system. PV panels often can be integrated into the roof itself, and some modules are actually designed as three-tab shingles or raised-seam metal roof sections. One benefit of these systems is their ability to offset the cost of roof materials.

How big should your PV system be, and what features should it have?

To begin, consider what portion of your current electricity needs you would like your PV system to meet. For example, suppose that you would like to meet 50% of your electricity needs with your PV system. You could work with your PV provider to examine past electric bills and

6

60

determine the size of the PV system needed to achieve that goal.

You can contact your utility and request the total electricity usage, measured in kilowatt-hours, for your household or business over the past 12 months (or consult your electric bills if you save them). Ask your PV provider how much your new PV system will produce per year (also measured in kilowatt-hours) and compare that number to your annual electricity usage (called demand) to get an idea of how much you will save. In the next section, we'll provide more information on estimating how much you will save.

Some solar rebate programs are capped at a certain dollar amount. Therefore, a solar electric system that matches this cap maximizes the benefit of the solar rebate.

To qualify for net metering in some service territories, your PV system must have a peak generating capacity that is typically not more 10 kilowatts (10,000 watts), although this peak may differ from state to state. Also, utilities have different provisions for buying excess electricity

produced by your system on an annual basis (see the section on net metering). Finally, customers eligible for net metering vary from utility to utility; for example, net metering could be allowed for residential customers only, commercial customers only, or both.

One optional feature to consider is a battery system to provide energy storage (for stand-alone systems) or backup power in case of a utility power outage (for grid-connected systems). Batteries add value to your system, but at an increased price.

As a rule, the cost per kilowatt-hour goes down as you increase the size of the system. For example, many inverters are sized for systems up to 5 kilowatts, so even if your PV array is smaller (say, 3 kilowatts), you may have to buy the same size of inverter. Labor costs for a small system may be nearly as much as those for a large system, so you are likely to get a better price for installing a 2-kilowatt system all at once, rather than installing 1 kilowatt each year for two years.

Roof Area Needed in Square Feet (shown in Bold Type)							
PV Module Efficiency (%)	PV Capacity Rating (Watts)						
	100	250	500	1,000	2,000	4,000	10,000
4	30	75	150	300	600	1,200	3,000
8	15	38	75	150	300	600	1,500
12	10	25	50	100	200	400	1,000
16	8	20	40	80	160	320	800

For example, to generate 2,000 watts from a 12%-efficient system, you need 200 square feet of roof area.

7

Printed by Books on Demand GmbH, Norderstedt / Germany